Key Skills Level 1:
of Number

Written to the 2004 Standards

Liam Gabrielle

Series Editor
Roslyn Whitley Willis

Published by
Lexden Publishing Ltd
www.lexden-publishing.co.uk

To ensure that your book is up to date visit:
www.lexden-publishing.co.uk/keyskills/update.htm

Acknowledgements

Special thanks for their tireless support and patience go to my wife and family and to Roslyn Whitley Willis, who was always willing to give direction in this mammoth task from the very start of an idea – *Liam Gabrielle.*

First Published in 2008 by Lexden Publishing Ltd.

Cover photograph of juggling balls by kind permission of Marcel Hol ©

British Library Cataloguing in Publication Data.

A CIP record of this book is available from the British Library.

ISBN: 978-1-904995-29-6

Typeset and designed by Lexden Publishing Ltd

Printed by Lightning Source.

Lexden Publishing Ltd
Email: info@lexden-publishing.co.uk
www.lexden-publishing.co.uk

PREFACE

The material in this book gives you the opportunity to understand Key Skills and practise them so you are able to meet the high standards set out in the Level 1 Key Skills Standards for Application of Number.

The introductory section of this book explains each of the Key Skills and how to gain a qualification.

This book is further divided into three distinct parts:

1   Reference Sheets

This section provides all the necessary background information to prepare you for Level 1 Application of Number. It provides useful exercises that will:

    aid your learning;

    can be used for revision; and

    prepare and aid you for the Part A Tasks and End Assessment questions.

2   Part A Practice Tasks

Working through these will help you produce work at the right level and prepare you for the End Assessment.

As you complete each task you will become more confident about what is expected in Key Skills and be able to use your knowledge and understanding to pass the End Assessment and put together a Portfolio of Evidence.

3   End Assessment Questions

This section provides examples of the type of questions that are likely to appear on an End Assessment paper and that you may have to pass as part of your Key Skills qualification.

Further resources

When your tutor thinks you have enough knowledge of Key Skills, she/he will give you an assignment, or assignments, to complete. Working successfully through the assignment(s) will show you are able to apply your knowledge and understanding, and produce work that will go into your Key Skills Portfolio of Evidence. These assignments are contained in the *Tutor's Resource* cd.

Additional resources and information can be found at www.lexden-publishing.co.uk/keyskills.

WHAT ARE KEY SKILLS? – A STUDENT'S GUIDE

Key Skills are important for everything you do, at school, at college, at work and at home. They will help you in your vocational studies and prepare you for the skills you will use in education and training and the work you will do in the future.

Key Skills are at the centre of your learning, and the work in this book provides you with the opportunity to develop and practise the Key Skill of Application of Number, through a variety of tasks. Having Key Skills Application of Number knowledge will help you produce work for other Key Skills and other tasks which occur in your studies.

There are six Key Skills

Communication is about writing and speaking.

It will help you develop your skills in:

- speaking;
- listening;
- researching;
- reading;
- writing;
- presenting information in the form of text and images, including diagrams, charts and graphs.

Application of Number is about numbers.

It will help you develop your skills in:

- collecting information;
- Interpreting information;
- carrying out calculations;
- understanding the results of your calculations;
- presenting your findings in a variety of ways, such as diagrams, charts and graphs.

Information and Communication Technology is about communicating using ICT.

It will help you develop your skills in using computers to:

- find and store information;
- produce information using text and images and numbers;
- develop your presentation of documents;
- communicate information to other people.

Improving Own Learning and Performance is about planning and reviewing your work.

It will help you develop your skills in:

- setting targets;
- setting deadlines;
- following your action plan of targets and deadlines;
- reviewing your progress;
- reviewing your achievements;
- identifying your strengths and weaknesses.

Problem Solving is about understanding and solving problems.

It will help you develop your skills in:

- identifying the problem;
- coming up with solutions to the problem;
- selecting ways of tackling the problem;
- planning what you need to do to solve the problem;
- following your plan;
- deciding if you have solved the problem;
- reviewing your problem solving techniques.

Working with Others is about working effectively with other people and giving support to them.

It will help you develop your skills in:

- working with another, or several, person(s);
- deciding on the roles and responsibilities of each person;
- putting together an action plan of targets and responsibilities;
- carrying out your responsibilities;
- supporting other members of the group;
- reviewing progress;
- reviewing your achievements;
- identifying the strengths and weaknesses of working with other people.

HOW TO GAIN A KEY SKILLS QUALIFICATION

Mandatory Key Skills

Communication
Application of Number
ICT

Practise Part A Key
Skills tasks in this book
to help you:

**Put together a
Portfolio of Evidence**

Usually an assignment written for
you by your tutor to cover the Key
Skill, or a number of Key Skills in
one piece of work.

The Portfolio is based on Part B of
the Key Skill Standards.

Pass a test –
called an **End Assessment**

This test is 40 multiple-choice
questions and, at Levels 1 and 2,
you have either 1 hour or 1 hour 15
minutes to complete it depending on
the Key Skill.

The questions are based on Part A of
the Key Skill Standards.

Good News!

If you already have some GCSE or ICT
qualifications, it may not be necessary
for you to take the End Assessment! Your
tutor will help you with this – it is called
PROXY QUALIFICATIONS.

Wider Key Skills

Improving Own Learning and Performance
Problem Solving
Working with Others

Put together a **Portfolio of Evidence**

Usually included in an assignment written for you by
your tutor to cover the Key Skills of Communication,
Application of Number or ICT.

The Portfolio is based on Part B of the Key Skill
Standards.

Practise Part A Key Skills
tasks in your vocational
studies to help you:

Opportunities to work towards achieving the Wider Key Skills are provided in the Portfolio assignment work and
are included in the *Tutor's Resource* that accompanies this text.

THE PORTFOLIO

STEP 1

Once your tutor has assessed your assignment work and you have passed, you will put your work into your portfolio.

A **Portfolio of Evidence** usually takes the form of a lever arch file with a **Portfolio Front Sheet** that shows:

 where you are studying;

 which course you are studying;

 which Key Skill(s) are in the portfolio;

 when you passed your End Assessment(s); and

 details of any Proxy Qualifications.

STEP 2

It is important to number every page of the work you put in your portfolio. This helps you complete the **Log Book** that your tutor will give you.

STEP 3

Complete the Log Book. This indicates where your evidence is to be found and also describes what is in the portfolio.

STEP 4

Check your Log Book entries carefully, making sure everything is correct and neat.

Get your tutor to check you have put your Portfolio together correctly.

STEP 5

Sign the Log Book and get the person who assessed your work to sign too.

Once you have completed your Portfolio of Evidence it is shown to someone outside your centre whose job it is to check it meets the Key Skills Standards. If this person agrees that it does, then you have **passed your Portfolio of Evidence**.

Application of Number

At **Level 1**, a learner should be able to use combinations of: **addition**, **subtraction**, **multiplication** and **division** with virtually no problems.

However, at Level 1, Application of Number tasks may require breaking down into manageable parts to form a solution. For example, using addition followed by division to form a solution. Combinations may vary.

The following reference sheets provide opportunities for you to review and practise the numeracy functions a learner needs for key skills.

It must be highlighted again that numeracy tasks can require a combination of functions, and as a learner you should be able to select appropriate methods to obtain answers.

UNDERSTANDING NUMBERS

Whole numbers

Read, write, order, positive/negative, estimate, compare, use.

Word	Number
one	1
ten	10
hundred	100
thousand	1,000
ten thousand	10,000
hundred thousand	100,000
million	1,000,000

A lottery winner won **£2,456,125.19** – this could be written or read as two million, four hundred and fifty-six thousand, one hundred and twenty-five pounds and nineteen pence. Remember to start with the largest number through to the smallest number.

There is also the **ordering** of numbers to consider, for example:

Numbers	
1st	first
2nd	second
25th	twenty-fifth
50th	fiftieth
100th	one hundredth
500th	five hundredth

Other examples can be dates such as *23rd May 2001*.

Recognising positive/negative numbers

Positive numbers and negative numbers can be such things as currency, temperature, time/date or units of measure, etc.

The simplest way to understand positive and negative numbers is to use a **number line**. Negative numbers always have a minus sign (-) before them, i.e. **-5° C** or **-£15**. Positive numbers don't usually display the + sign before them. Zero is always in the middle. For example:

Temperature

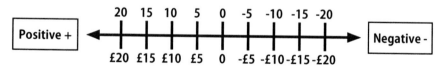

Bank balance

The above diagram shows both temperature and a bank balance for comparison.

A numberline can also be vertical. The choice is yours.

Rules for adding and subtracting

Adding a negative number is the same as subtracting. For example:

$7 + (-4)$ is the same as $7 - 4 = 3$

Subtracting a negative number is the same as adding. Two negatives make a positive. For example:

$(-5) - (-3)$ is the same as $(-5) + 3 = -2$

Q A fridge has a temperature of **2°C** and a freezer has a temperature of **-9°C**. What is the difference in temperature?

✓ Think of, or draw a numberline and work out the difference. Equals **11°C**.

$$\boxed{11°C}$$

10 9 8 7 6 5 4 3 2 1 0 -1 -2 -3 -4 -5 -6 -7 -8 -9 -10

Q Inside an aeroplane the temperature is **22°C** but outside it is **-40°C**. What is the difference or spread of temperature?

✓ There is **22°C** on the **positive** side and **40°C** on the **negative** side. Add the two temperatures together $(22 + 40)$ to get a difference of **62°C**.

Q A business has an income during a month of £100,000. However, costs were £50,000 and capital outlay was £200,000. How much money does the business have this month?

✓ First, separate income (which can be thought of as profit) from costs and capital outlay (which can be thought of as expenditure). Then treat it as a simple minus sum.

So, £100,000 is income or **profit**. Expenses (**loss**) were costs of £50,000 and capital outlay of £200,000, add these together to get £250,000.

Finally, £100,000 **minus** £250,000 will leave **-£150,000**.

Estimation

Estimation is using your best judgement or approximation.

You can use **estimation** to say how much fluid is in this beaker:

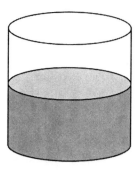

You can estimate the beaker as being **half full**. It can be also be said that it is **half empty**.

Look at this tape measure:

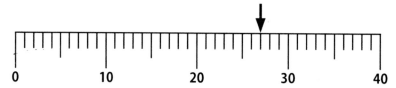

After looking at this tape measure it can be estimated that it is **approximately** 25 units. It can also be said it is approximately 30 units.

Q Three items from a shop cost £31.80, £10.05 and £4.99. Estimate the total cost?

✓ Simplify the amounts to the nearest **whole pound**. Then calculate an answer:

£32 + £10 + £5 = £47

Q A worker charges £30.20 per hour and takes 21 hours to complete it. Estimate the final bill?

✓ Make **£30.20 = £30** and **21 hours = 20**. So £30 multiplied by 20 = **£600 final bill**.

Q A boy has maths homework to do. He has 4 rows of problems.
There are 9 problems in each row. If he can do 10 problems every 15 minutes, estimate how long will it take for him to finish his homework?

✓ You can estimate **9 as 10**, and 4 rows by 10 problems = **40 problems**. So 10 problems every 15 minutes would mean **15 x 4** = 60mins. Answer = 1hr for his homework.

Rounding off

Rounding off is a way of simplifying numbers. For example, if you won **£3,673,017.12** it would be easier to write **£3,500,000** or **three and a half million pounds**. Rounding off numbers also makes it easier if you want to estimate. Rounding numbers to the nearest 10, for example, means finding which 10 they are nearest to. Rounding a number to the nearest hundred, to the nearest thousand or even bigger can be done in the same way.

Rules for rounding off

If you were asked to round off **145** to the nearest **10** it would either become **140** or **150**. Look at the following rules:

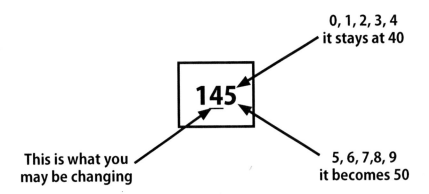

Q   A pencil is 27cm long. How long is it to the nearest **10cm**?

✓   You are looking to possibly change the **10cm** unit. Look at the **7 unit** beside the **20 unit** as this will affect changing this up one unit or staying the same. As it is 7 we must round up to **30cm**.

Q   A sprinter runs 1,400 metres around a track. What is this to the nearest one **thousand** metres?

✓   You are looking to possibly change the **1,000m** unit. Look at the **4** beside the one thousand metre unit. This means that the one thousand metre unit will not change, so the answer is **1,000m**.

Accuracy

Accuracy is used to define an acceptable level. For example, if you are completing a long multiplication sum you may be required to give your answer to **two decimal places accuracy**. Your answer may then be **19.23** to two decimal places.

In some respects it is **similar** to rounding off as it also uses given levels of accuracy.

Other examples of where accuracy may be used are in **measuring** and **time**.

You may **measure** the length of a room and use an accuracy of one centimetre rather than being precise to the exact millimetre. But if you were measuring to fit a new door frame an accuracy of one millimetre would then be necessary.

If somebody asked you for the time a usual reply might be, "nine thirty" or "twenty to eleven". Giving the time to the nearest second would generally not be used. However, a more precise and accurate time would be acceptable in a 100 metres race where the difference of one hundredth of a second is crucial.

Ratio

A ratio is a way of proportioning numbers or quantities into **parts**. To put it another way, ratios are about **sharing items out**. It can be as simple as mixing paint colours such as, red and black to get another colour.

Ratio is also another way of comparing the relationships between two or more quantities such as in scale diagrams.

At Level 1, you should be able to work out ratios of 2 parts and recognise that ratio can be displayed as: **1 to 20, 1 : 20, 1/20**.

Mixing paint is an ideal example of ratio. Consider the following:

In this mixing pot of paint there are **2 parts of** one colour and there are **3 parts of another colour**.

Importantly, you must remember that there is a total of 5 parts.

If we were to make up **5 Litres** of paint we would need **2 Litres** of one colour and **3 Litres** of another colour. However, it is usually not that easy.

If we require **20 Litres** of paint with the ratio **2 : 3** yellow to blue, how much of each colour will we need?

First, we add the **2 + 3 = 5** as this shows we have **5 parts** in **total**. Next we divide 5 in to 20 Litres as this will show the size of **one part** of the 5. So **20 ÷ 5 = 4**.

Next we use this answer of 4 and multiply it to each colour:

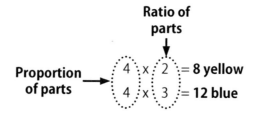

Let's use a check calculation to **prove** this works by adding 8 + 12 = 20. **So we have 8 Litres of yellow to 12 Litres of blue (2 : 3).**

Another example would be to divide an amount of £400 in to a ratio of **5 : 3**

First add 5 + 3 = 8 (parts) then

£400 ÷ 8 = £50 (for 1 part of the total)

So, £50 x 5 = £250 (5 parts of total)

£50 x 3 = £150 (3 parts of total)

£400 in a ratio of 5 : 3 = £250 to £150 (check calculation 250 + 150 = 400).

 There are 2,000 employees in a ratio of **3 : 7 males to females**. How many males are there?

 Add 3 + 7 = 10 Next divide 2,000 by 10 equals 200 (**1 part**) Then 3 (males) x 200 = 600. There are **600 males.**

Scale

Scales use ratio in a similar way. If you photocopy a picture by **1 : 1** it will come out the **same size**. But if you were to photocopy it at **1 : 2** it will come out **half the size (1/2)**. However, you may want to enlarge the picture by **2 x**. This will be a scale of **2 : 1** making the picture twice as big. For example:

 1 : 1    **2 : 1**   **1 : 2**

At **Level 1** you should be able to use and understand scales in things like diagrams or maps.

Q   If you made a model boat to the scale of **1 : 22** and the original was **40 metres** long, how big would the model be?

✔   First, remember that the model will be **22 times smaller**. So divide 40 by 22, 40 ÷ 22 = 1.82m, to an accuracy of 2 decimal places.

Q   A carpenter uses a scale of **1cm to 2m** for a kitchen plan. How big would a kitchen unit be if shown on the drawing as **1.5cm by 0.5cm**?

✔   Since **1cm = 2m**, multiply **1.5 by 2** (1.5 x 2) this equals **3**. Next multiply **0.5 x 2** equals **1**. The life size dimensions of the kitchen unit will be **3m by 1m**.

Q   Look at the map. What is the distance from Uptown to Weston using a scale of **1cm to 0.5 miles**?

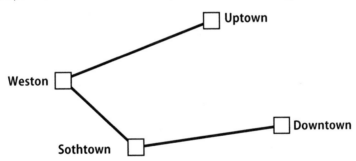

✔   You would be provided with a ruler to measure the distance between Uptown and Weston. **For every centimetre the distance is equal to 0.5 miles in real life**.

Proportion

Proportion is when two or more quantities increase at the **same rate** as each other. For example, the further the distance that you travel in a car, the more fuel you are going to use. From this you can see a **direct relationship**.

Other examples include – the more hours you work in employment the more money you will be paid in wages. Note also, the more you earn the more in tax you are likely to pay.

The reverse is also true and is known as **inverse proportion**. For example, it takes two men six days to paint a house. If there were four men painting then it should take three days to paint the house.

 An engineering company employs **10 people to produce 500** car parts over **10** days. How long would it take 15 people to complete the parts? ㄱ ⊆ ○

 The people employed to produce the 500 parts has been increased by **50% (1/2)**, so the time to produce the parts should be reduced by the same amount. So **half** (50%) of 10 days equals 5 days to produce the parts. ㄱ ⊆ ○

Fractions

Understand, order, increase/decrease, compare.

Fractions are **parts of a whole**. A **half** (1/2) of **one** (1) is a **half** (1/2), and **two halves make one** (1/2 + 1/2).

At **Level 1** you should be able to use and understand simple fractions, and also see their relationship to decimals and percentages.

Fractions are made up of **two parts**, a numerator and a denominator. Below is a diagram to show the relationship between them:

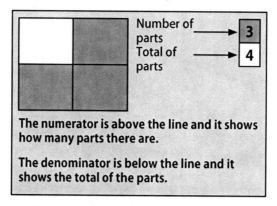

It is important to know the size and relationship fractions have to each other. The following chart shows some fractions as an **image**; the way a fraction is **written** and **read**. Included within the **Read** section is an explanation.

Model	Fraction	Read
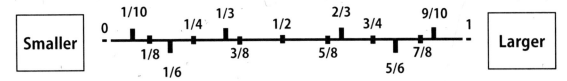	$\frac{1}{2}$	One half, one out of two, one divided by two.
	$\frac{1}{4}$	One fourth, one out of four, one divided by four.
	$\frac{2}{3}$	Two thirds, two out of three, two divided by three.
	$\frac{3}{5}$	Three fifths, three out of five, three divided by five.

It may be easier to think of fractions by using a **numberline**. Here is an example using some popular fractions:

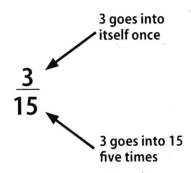

Fractions sometimes need to be **simplified**. For instance **2/10** can be simplified because 2 goes into 10 five times and into itself once, giving an answer of **1/5**. Here is another example:

3 goes into itself once

$$\frac{3}{15}$$

3 goes into 15 five times

Answer = **1/5**

Q What is **3/4 of £2**?

First divide the quantity **£2** by the **denominator** of the fraction; 2 ÷ 4 = 0.5, then multiply the answer by the **numerator 3**. This is 3 x 0.5 = 1.5. So **3/4 of £2 equals £1.50.**

Q A man is paid, after taxes, £180 in wages. He estimates that **1/3** was deducted in tax. What was his full wage before tax deductions?

If **2/3 equals £180** (remember he's lost 1/3 in tax). Divide 180 by the **numerator 2**, equals 90. 90 is the multiplied by the **denominator 3**. 3 x 90 = 270. The wage before deductions was **£270.**

Decimals

Understand, order, increase/decrease, compare.

A decimal is a number that uses place values and a decimal point to show amounts that are **more** than or **less** than one, such as zero point two five (**0.25**) or **£5.95**.

At **Level 1** you should be able to work with decimals using basic arithmetic and compare/convert to fractions and percentages.

Importantly, you must be able to order decimals from **largest** to **smallest**. For example:

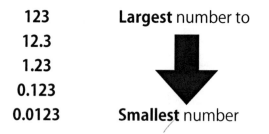

123	**Largest** number to
12.3	
1.23	
0.123	
0.0123	**Smallest** number

You may want to use a **numberline** to help order decimals, such as this one:

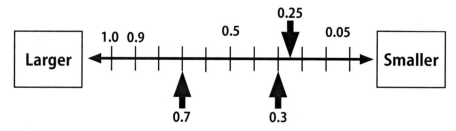

Q Place the following decimals in to order **largest to smallest**:

2.9
1.9
1.09
2.09

First look to the **left** of the decimal place and decide which is the **largest number**. Then from the **right** of the **decimal place** in the **first column**, decide which is the next largest. Use the same technique again moving to the right from the previous column. Placed in order of size the decimals will be:

2.9, 2.09, 1.9, 1.09

Q How do you multiply 3.75 x 9.2?

This should be treated the same as a **long multiplication** sum first. Look at the following method:

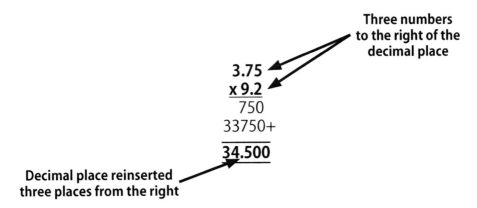

Three numbers
to the right of the
decimal place

3.75
x 9.2
750
33750+
──────
34.500

Decimal place reinserted
three places from the right

After the multiplication sum is **complete** the decimal place is **reinserted**. This is done by **adding** up the number of figures used to the right of the decimal place. **In this case there are three.** The decimal point is placed back in **three places from the right** in the final answer as shown.

Percentages

Understand, order, increase/decrease, compare.

At **Level 1** you will be expected to calculate and use percentage values, and also convert to fractions and decimals.

Percent means **out of a 100**. It is a set of number values usually up to a maximum value of **100%**. A good way to show percentage values is to use a **numberline**. Here is an example:

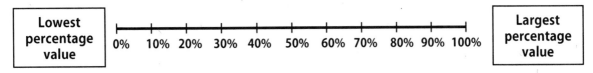

| Lowest percentage value | | Largest percentage value |

0% 10% 20% 30% 40% 50% 60% 70% 80% 90% 100%

100% means the **full value** of a figure. **50%** means **half** of the value, for example a music CD normally selling for £12 would be £6 if it was on sale at **50% off**.

But how could you work out 50% off on a CD costing £9.72?

The usual mathematical method would be:

$$\frac{£9.72}{100} \times 50 = £4.86$$

Another method is to multiply 50 x £9.72 = 486 Then divide the answer by 100, which equals £4.86.

And another method is to find 10% of £9.72 by moving the numbers one place to the right of the decimal point = 0.972. Next, multiply this answer by 5 = £4.86.

These are different methods that give the same answer. The method used is your choice.

Q At a sporting event there are a **total of 1,200** people. **300** of these were female spectators. **What is the number of female spectators as a percentage of the 1,200?**

✓ If you were using a calculator; then **300 ÷ 1,200 x 100 = 25%**
Another method would be to realise that 1,200 is the full number and equals 100%. If you halve 1,200 (**which is the 100% figure**) it would equal 600 (**which is 50%**), this can be halved again, 600 (**50%**) would equal 300 (**which is 25%**). **The females would equal 25% of the total.**

Fractions, decimals and percentages

As a **Level 1** student you must be able to understand the **relationship** between fractions, decimals and percentages. Again a **numberline** is useful to note this relationship. Here is an example:

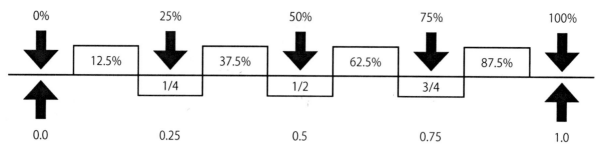

Q   **Forty** athletes are trying to get on the national team. **25%** initially fail the first trial. At the second trial **1/3** fail. Then at the final trials, only **0.5** pass. **How many athletes make it on to the national team?**

✓   From the original **forty** 25% fail, this leaves us with **30** because 25% of 40 = 10 (**4 x 10 = 40**). At the second trials **1/3** of thirty now fail, this leaves us with **twenty (1/3 of 30 = 10, or 3 x 10 = 30)**. Now at the final trials only **0.5** pass from **20**, this leaves us with ten (**0.5 x 20 = 10**). So **there are only 10 who make the national team**.

Q   What is **3/8** as a **decimal**?

✓   To convert **3/8** in to a decimal **divide 3** (numerator) **by 8** (denominator) 3 ÷ 8 = 0.375 (**here is the long division working out**) you may work it out differently though.

$$
\begin{array}{r}
0.375 \\
8\,\overline{)\,3.000} \\
24 \\
60 \\
56 \\
40 \\
\end{array}
$$

Q   What is **0.25** as a **fraction**?

✓   Decimals are easily converted into fractions. **0.25 is a quarter of 1** or in other words, there are four 0.25s in 1. So **0.25 = 1/4**. Using a numberline of decimals and fractions is useful here.

Q   What is **20%** as a **fraction**?

✓   Again possibly using a numberline may help to see the relationship of percentages and fractions. Alternatively, put the **20%** over the **100%** as this is the fraction **20/100**. Now the fraction needs to be simplified. **20 goes into itself once and into a 100 five times giving an answer of 1/5.**

Using a calculator

At **Level 1** you should be able to use a calculator confidently, using the basic arithmetic functions, also simple percentages, decimals, negative numbers and fractions in a range of calculations.

You are **not** allowed to use a calculator during the End Assessment. However, you are encouraged to check and use a calculator during Part A Tasks and when completing your portfolio. **But you should not rely on a calculator for every task.**

UNITS OF MEASURE

Understand units of measure, order, increase/decrease, compare.

The **metric system** is a system of measurement that describes the measurement of an object, its weight, or its volume. It uses the decimal system in powers of ten. The units are measured with rulers, tapes, scales, etc.

There is also the **imperial measuring system**. This is still occasionally used today in certain areas of employment and examples are given on *page 18*.

At **Level 1** you will be expected to be able to use the metric system and its various units. You will also be able to convert or compare between the imperial as necessary.

The metric system

Length is measured using the units of:

	Millimetre (mm)
10mm = 1cm	Centimetre (cm)
100cm = 1m	Metre (m)
1,000m = 1km	Kilometre (km)

The following are examples of measurements in length:

8mm =	0.8cm
12mm =	1.2cm
435mm =	43.5cm
37cm =	0.37m
87.5cm =	0.875m
490m =	0.49km
3,590m =	3.59km

Weight is measured using the metric units of:

	Grams (g)
1,000g =	1 Kilogram (kg)
1,000kg =	1 Tonne

The following are examples of weight:

250g =	0.25kg
2,755g =	2.755kg
1.25kg =	1,250g
4.8kg =	4,800g

Capacity is measured using the units of:

	Millilitre (ml)
10ml =	Centilitre (cl)
100cl =	Litre (L)
1,000ml =	Litre (L)

The following are examples of capacity:

$$
\begin{aligned}
30ml &= 3cl \\
200ml &= 20cl \\
750ml &= 0.75L \\
5,500ml &= 5.5L
\end{aligned}
$$

The imperial system

The imperial system of measuring uses its own units of measurement.

At **Level 1** you should understand the imperial system and convert to the metric system with given formula if necessary.

Length in the imperial system:

$$
\begin{aligned}
& \text{Inches (")} \\
12\ \text{inches} &= 1\ \text{foot (ft)} \\
3\ \text{feet} &= 1\ \text{yard} \\
1,760\ \text{yards} &= 1\ \text{mile}
\end{aligned}
$$

Examples of length:

4 inches

1ft 3 inches

1½ miles

Weight using the imperial system:

$$
\begin{aligned}
& \text{Ounces (oz)} \\
16oz &= 1\ \text{pound (lb)} \\
14lb &= 1\ \text{stone (st)} \\
2,240lb &= 1\ \text{ton}
\end{aligned}
$$

Examples of weight:

6lb

2½lb

4¼st

Capacity in the imperial system:

$$
\begin{aligned}
20fl\ oz &\quad 1\ \text{pint} \\
8\ \text{pints} &= 1\ \text{gallon}
\end{aligned}
$$

Examples of capacity:

4 pints

9 gallons

Measurements in the **metric** and **imperial** system can be taken using various instruments such as scales, tapes, jugs, digital appliances etc. At **Level 1**, you will be expected to read these instruments accurately, and to given levels of accuracy, in what ever system it is measuring.

The following is an example of a scale using both **metric** and **imperial** systems:

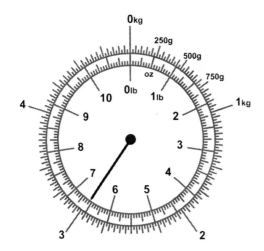

This scale shows the various weight scales in comparison to each other.

The weight is slightly over 6 1/2lb and just under 3kg.

Time

Time is measured in different units and can be measured either **digitally** or by an **analogue** clock using the **12-hour** or **24-hour** format.

At **Level 1** you should be able to use the **12/24-hour** format and work using seconds, minutes, hours and days.

Example of time units used:

60 seconds =	1 minute (min)
60 minutes =	1 hour (hr)
24 hours =	1 day
7 days =	1 week (wk)
12 months =	1 year (yr)
1 year =	365 days

The following are examples of time used daily:

1 min 30 secs

4 hrs 45 mins

48 hrs

2 days 9 hrs

Here is an example of a digital clock showing 19:05 hours. It shows the hours and minutes. The **am** and **pm** may also be shown:

This can also be displayed as **7:05pm** in the 12-hour format.

Here is an example of an analogue clock showing **10:10am**. It could also be showing **22:10** in the 24-hour format:

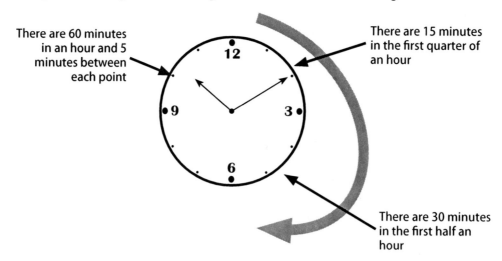

There are 60 minutes in an hour and 5 minutes between each point

There are 15 minutes in the first quarter of an hour

There are 30 minutes in the first half an hour

Q An office worker takes approximately **5 minutes** to type **140 words**. How long will it take for the worker to type **4,200 words**?

✓ 4,200 words **divided** by 140 words, this equals 30. Then 5 minutes multiplied by 30 equals 150 minutes. **This can also be written as 2 hrs 30 mins.**

Q The ferry crossing from England to Ireland leaves at **22:30 hrs** and takes **2hrs 45 mins**. What time will it arrive?

✓ First **add 2 hrs** on to **22:30 hrs**, equals **00:30 hrs**. Then add the remaining **45 mins** to this, equals **01:15 hrs**. It may help to draw a clock face in more complex problems to aid finding the answer.

Temperature

Temperature is generally measured in **degrees Celsius** and shown by the symbol **°C**. There is also another measure of temperature and that is in **degrees Fahrenheit** and this is shown by the symbol **°F**. Degrees Celsius tend to be the most widely used.

At **Level 1** you will be expected to read both measuring systems. There is a formula to convert between temperatures, but this is not required at Level 1.

Temperature can be measured by a thermometer or digitally. It is important to realise that a temperature may go **below 0** and this makes it a negative number. For example, it can be said that the temperature outside is **5°C below 0**. This is the same as **-5°C**.

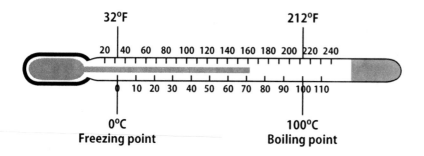

32°F   212°F

0°C
Freezing point

100°C
Boiling point

This is an example of a thermometer showing both **Celsius** and **Fahrenheit**.

Notice that the scales are different. Boiling point is 100°C or 212°F.

Using a digital scale temperature will look like this:

37.5°C

Q The temperature has **dropped** five degrees during the night from **3°C** during the day. What is the temperature during the night?

✓ Using a **numberline** here is useful. Work backwards from

$$\xrightarrow{\text{5}}$$
3, 2, 1, 0, -1, -2

and we get an answer of **-2°C**.

SHAPE AND SPACE

Find area, perimeter and volume.

At **Level 1** you should be able to work out area, perimeter and volume in simple shapes.

Area

Area is calculated by length multiplied by width (**L x W**) and shown by the symbol ^{2} (**squared**). **This is actually a simple formula where letters are replaced by numbers**.

At **Level 1** area calculations are generally based on square, oblong and simple **L shaped** dimensions. But the process may involve working out missing dimensions to find the area. For example, look at the storeroom illustrated below:

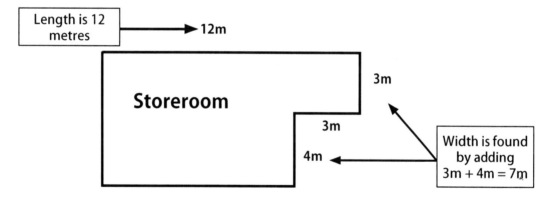

The **general area** can be found by multiplying 12m x 7m = 84m^2. But this is not the correct area. There is the **cut out** of 4m x 3m = 12m^2 to be taken away from the general area total. So **84m – 12m = 72m^2** for the area of the **L shaped** storeroom. *See Diagrams, page 27 for a further example.*

Sometimes you may have to find the area of circles. The formula used is always given including the value to use for π **and this is nearly always 3**. For example:

Area of a circle = πr^2

$\pi = 3$

r = radius

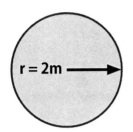

$Area = \pi r^2$

First, put the numbers into the formula

$Area = 3 \times 2m^2$ (remembering to multiply the squared value first)

The circle area = $12m^2$

Q What is the area of this house shown by the plan view?

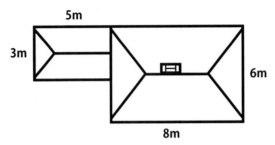

✔ There are **two** simple **oblong** areas to work out then add together to get a total. So the first one is **3m x 5m = 15m²** then, **8m x 6m = 48m²**. Then add the two together to get the **total area** **15m² + 48m² = 63m² total area.**

Perimeter

Perimeter is the distance **around an object** and is found by adding all the dimensions together to form a **total**. If we were to find the perimeter in this example we would have to work out the **missing** dimensions first.

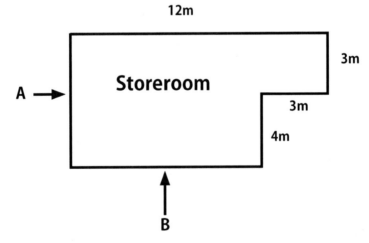

The value of **A** is found by adding the dimensions 3m and 4m together (**7m**) as they are **parallel** with A. Then the value of **B** is found by taking 3m away from 12m (**9m**). So the **perimeter is 12m + 3m + 3m + 4m + 9m + 7m = 38m.**

Volume

The volume of objects is generally found by multiplying the **length x width x depth** and is shown by the symbol ^{3}.

At **Level 1** you are expected to work out simple volume in box shapes. All of the dimensions are given.

 What is the volume of the following object?

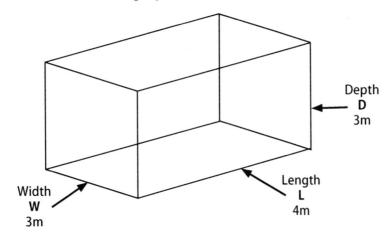

 The volume is found by **L x W x D**. Replace the letters with the numbers and this is 4 x 3 x 3 = 36m^3.

Using formula

At **Level 1** you will not be given specific formula to do. However, you will use formula to find the **area** and **volume** in simple objects.

Formula is something as simple as finding area, **L x W** where **L** is length multiplied by **W** width. It can also be in the volume of a cube, **L x W x D**. It is also used in finding the area of a circle by πr^2 or in the volume of a cylinder, $\pi r^2 d$, *but these are expected at Level 2.*

Whatever formula is given for you to use, it is simply a case of replacing **letters** or **symbols** with **numerical values**.

DATA

At **Level 1** you will be expected to work with up to **ten** sets of data displayed in various forms. You will need to understand, read, write, order, and compare data in tables, charts, diagrams and graphs.

Data can be displayed in not only numerical form, i.e. numbers, but also in various graphical modes. In other words, displayed in images, tables, charts, diagrams and graphs. This makes it possible to show comparisons, trends or even patterns in data.

Tables

A table is an easy way to record and display data. For instance, you have recorded the colour of cars in the college car park. There is a suitable title and column headings for colour and number of cars recorded. You may also place a total at the bottom to show how many cars there were in the car park.

Cars in the College	
Colour	**Number**
Silver	12
Blue	8
White	5
Black	10
Red	7

Tables can also display a vast amount of information and this may seem confusing at firstglance, but learn to study carefully and understand the information displayed.

Here is an example of a typical bank repayment table. If you were interested in borrowing say £1,000 to buy a car, you can compare figures for **12** months, **24** months, **36** months, **48** months or **60** months.

Bank Repayment Period					
Amount of Loan £s	**12 Months**	**24 Months**	**36 Months**	**48 Months**	**60 Months**
£10,000	£916.67	£504.17	£369.72	£305.02	£268.42
£7,000	£641.67	£352.92	£258.81	£213.51	£187.89
£6,500	£595.83	£327.71	£240.32	£198.26	£174.47
£6,000	£550.00	£302.50	£221.83	£183.01	£161.05
£5,500	£504.17	£277.29	£203.35	£167.76	£147.63
£5,000	£458.33	£252.08	£184.86	£152.51	£134.21
£4,500	£412.50	£226.88	£166.38	£137.26	£120.79
£4,000	£366.67	£201.67	£147.89	£122.01	£107.37
£3,500	£320.83	£176.46	£129.40	£106.76	£93.95
£3,000	£275.00	£151.25	£110.92	£91.51	£80.53
£2,500	£229.17	£126.04	£92.43	£76.26	£67.10
£2,000	£183.33	£100.83	£73.94	£61.00	£53.68
£1,500	£137.50	£75.63	£55.46	£45.75	£40.26
£1,000	**£91.67**	**£50.42**	**£36.97**	**£30.50**	**£26.84**

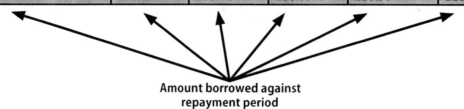

Amount borrowed against repayment period

Charts

Charts can be very appealing in appearance and can make viewing data easy, especially when comparing or contrasting. Rather than just showing numbers, they can be in the form of a **pie chart** or a **bar chart**, etc. Generally at **Level 1** the charts used tend to be either pie charts or bar (vertical, horizontal) charts.

Pie charts

Here is an example of a pie chart showing the ages of people attending a sports club. There is an appropriate title and labels for the segments:

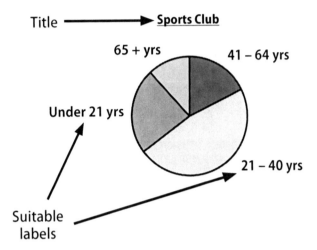

It can be clearly seen that the **most popular** age group is the 21 – 40 year olds.

The **least popular** group is the 65 years and over.

It is also easy to compare the age groups as a whole or individually.

However, the age groups can also have a numerical value applied of **figures or percentages and even degrees** displayed alongside the groups.

This adds more detail to the chart.

Bar charts

Bar charts can display single sets of data and also two or more sets of data. This is useful when **comparing information**. However, this is good as long as there are not too many bars displayed, as understanding the information can be confusing.

Bar charts tend to be displayed either **vertically** or **horizontally**. A title is generally included and axes displayed with suitable scales or labels.

Data

Here is an example of a vertical bar chart of when people learned to ride a bicycle:

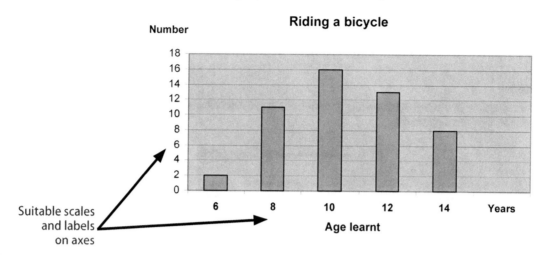

There is a title explaining the data (subject of the chart) and the horizontal axis shows the age learned. The vertical axis shows the number of people who learned. From the axes it can clearly be seen that **most** people learned to ride a bicycle at 10 years of age or that **only two** people learned to ride a bicycle at 6 years of age, etc.

Remember it is important to label the axes correctly, as failure to do so will lead to errors in data understanding.

Here is a vertical bar chart showing three sets of data over a three-month period. This is used when data needs to be compared. Importantly, there is a **key** identifying the separate bars on the graph.

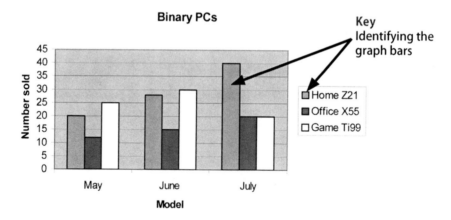

There is a title and suitable labels on the horizontal and vertical axes.

Diagrams

Diagrams are a good way of showing information such as the dimensions of a plan for an office. At **Level 1** you may have to work out missing dimensions and use these in your calculations of **area** or **perimeter**.

The following diagram has **missing dimensions**, but these can be worked out from other dimensions given, by either adding or subtracting values:

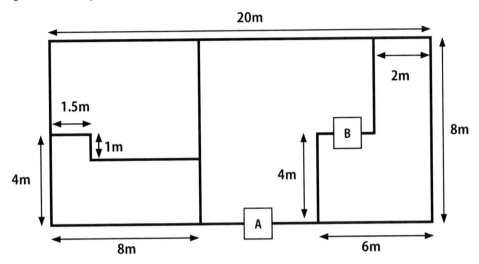

Dimension **A** can be found by adding the 8m and 6m either side together to get 14m and then taking 14m from 20m which is the opposite side. So **A** equals 6m.

B is found by taking 2m from 6m. Leaving **B** as 4m.

Diagrams may also have other images inserted into them to aid the display of information or problems. The following is an example used in a volume problem:

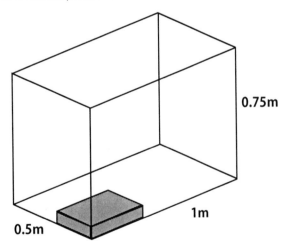

Q Using the following diagram, what are the missing dimensions?

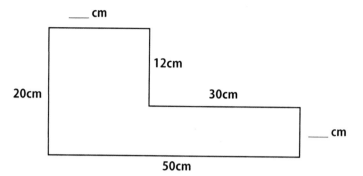

 Using the dimensions given, work out the **missing ones**. The top dimension is found by **50cm – 30cm = 20cm** and the side dimension is found by **20cm – 12cm = 8cm**. So the missing dimensions are **20cm** and **8cm**.

Graphs

Graphs are another useful way of displaying numerical information. Like bar charts they should have a **clear title** and **axes** using a **suitable scale**. Importantly, they must be labelled correctly. The following is a line graph displaying, the average temperature in degrees Celsius over five months in London:

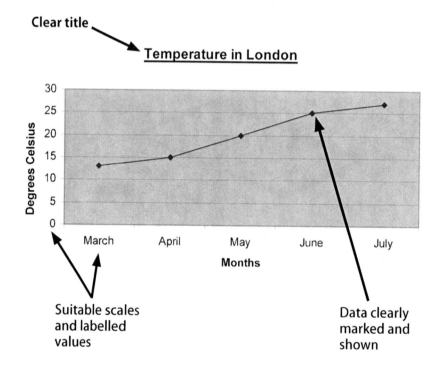

The line graph is **clearly titled** and there are **suitable scales** for the horizontal and vertical axes. The axes are also **correctly labelled** showing **clear values**.

Two sets of data
displayed

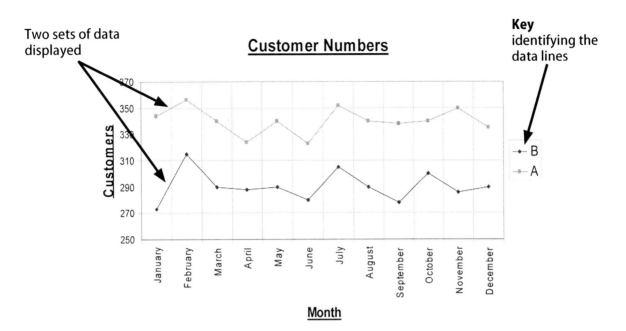

Customer Numbers

This is a **dual line graph** and it is used to display and compare **two sets of data**. There is a suitable title and the axes are clearly labelled, however there is the addition of a **key** to **identify each line**. The top line on the graph is **A** because it has a **box** on the plot points and **B** has **diamonds** on its plot points. **Notice that in this example, the key is upside down to the graph.**

Q   The following vertical bar graph is **missing** information. What is missing?

- International Airport -
Flights taking off late

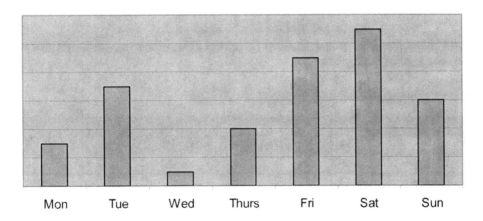

Looking at the bar graph the title looks correct as does the horizontal bottom axis. However, a scale and suitable label is **missing** from the **vertical** axis. So we don't know how many flights took off late.

Q From the horizontal bar graph how many people, **below** the age of **12**, learned to ride a bicycle?

Age learnt to ride a bicycle

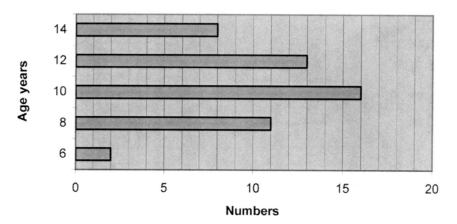

We are only interested in the numbers **below 12** years of age. That will be the 10 years, 8 years and 6 years numbers. They are **16+11+2 = 29** We have **29** who learned to ride a bike below 12 years of age.

Q From the table related to job types below, how many people have **nursing** jobs?

Job types	
Career	**Number**
Teacher	57
Doctor	15
Solicitor	12
Nurse	84
Engineer	57

From the table identify the career asked for **nursing** and read of the number of jobs. In this case there are **84 people with nursing jobs**.

Q From the following table, what is the gross profit in March?

Expenditure in £000s			
	January	**February**	**March**
Cash Sales	60.5	32.1	27.8
Callout Fees	1.5	2.8	3.1
Gross Profit	62	34.9	30.9
Expenditure			
Suppliers	20.3	10.2	0.9
Salaries	1.8	1.2	1.2
Bills	0.75	0.45	0.4
Totals	22.85	11.85	2.5

First of all it is important to note that the title states **expenditure in £000s**. This means that any figure in the table is actually hundreds of pounds. **So the gross profit for March is 30.9 £000s and this equals £30,900**.

Q From the line graph provided find how many **kilograms** there are in **55lb**?

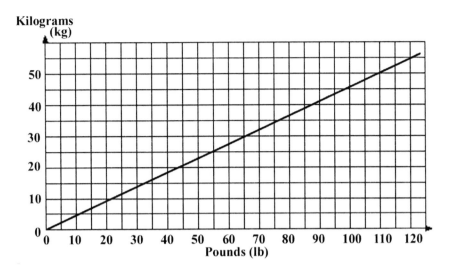

First, look along the **horizontal axis** as this shows **55lb**. Next, read up vertically to where **55lbs** intersects the diagonal line, then read off horizontally across on to the **vertical** kilogram scale. **This will show 25kg.**

Q Look at the dual line graph and identify the **least popular** month for **B** customers.

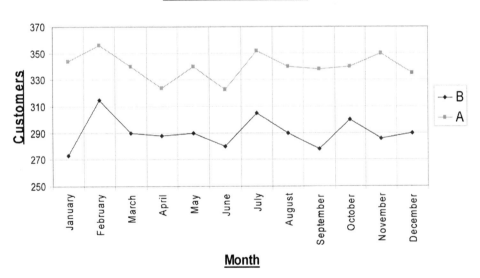

The first thing to do here is **identify** which is **B** by using the key on the side of the line graph. **B has the diamond shape** on the plotted line. Then we look for the lowest point. **This is January.**

STATISTICAL MEASURES

Understand, compare – **mean** and **range** in data in various forms.

Mean

Mean is another word for **average** and is treated in the same way. It can be used for finding the **average mark** in a set test or in **comparing** two sets of data.

If you have **ten** numbers and you wish to find the **mean**, first **add the ten numbers** together to **form a total**, and then **divide the total by ten** to find the mean number. **This works for any sets of numerical data.**

At **Level 1** you will be expected to work with at least ten sets of numerical data.

Q   What is the **mean** from the following test scores?

23	31	29	23	26
31	34	29	29	25

✓   Add the **ten** numbers together to form a total, **23+31+29+23+26+31+34+29+29+25 = 280**

Divide **280 ÷ 10 = 28** mean number.

Range

Range is used for finding the **spread** in sets of numerical data. It is calculated by finding the **largest figure** and taking away the **smallest (L – S)**. For example, if we were to look at a chart of temperatures and found **37°C** was the **largest** temperature and **7°C** the **smallest**, we would make the calculation as shown below:

37 minus 7 (37-7) = 30°C (**Range = 30°C**)

At **Level 1** you will be expected to work with at least ten sets of numerical data.

Q   A cyclist travels various distances over five days and records this information. What is the range?

Distance in kilometres
12.7, 17.9, 25.5, 11.2, 29.1

✓   **29.1** is the **largest. 11.2** is the **smallest.** So 29.1 minus 11.2 = **17.9 Km range.**

Q   The weekly wage of some office workers was recorded as £345.21, £210.85, £177.25, £421.21, £367.91, £119.23, £112.23, £211.84, £201.48. What was the range?

✓   **Largest** £421.21 – **smallest** £112.23 = **£308.98 range.**

APPLICATION OF NUMBER: PART A, PRACTICE TASKS

TASK DESCRIPTION GRID

Number and title	Page	Activities	Refer to reference sheet(s) on page(s)
1 Swift Feet	34	Whole numbers, statistics (mean), decimals, accuracy, range, tables, time.	6, 32, 14, 9, 24, 19
2 Stepping Stones	36	Units of measure, statistics, decimals, accuracy, whole numbers, temperature, weight, scales, graphs.	17, 32, 14, 9, 6, 20, 17, 18, 11, 28
3 Discounted	38	Units of measure, statistics (mean), fractions, area, whole numbers, accuracy, percentages.	17, 32, 12, 21, 6, 9, 15
4 Half Baked	40	Whole numbers, accuracy, decimals, area, volume, scales, ratio, units of measure, fractions, volume, diagrams.	6, 9, 14, 21, 23, 11, 10, 17, 12, 16, 23, 27
5 At a Fraction	42	Whole numbers, accuracy, fractions, area, units of measure, ratio.	6, 9, 16, 12, 21, 17, 10
6 What's the Point	43	Whole numbers, decimals, accuracy, rounding off, units of measure.	6, 14, 9, 17
7 Decorated Paint	44	Whole numbers, decimals, accuracy, rounding off, area, volume, scales, ratio, units of measure, diagrams.	6, 14, 9, 21, 11, 10, 17, 27
8 Well Displayed	47	Whole numbers, decimals, accuracy, statistics, tables, charts, graphs.	6, 14, 9, 32, 24, 25, 28
9 To Scale	50	Whole numbers, accuracy, decimals, scales, area, units of measure, proportion, diagrams.	6, 9, 14, 11, 21, 17, 12, 27
10 Mixing It Up	52	Whole numbers, decimals, fractions, percentages, diagrams.	6, 14, 16, 27
11 One Stop Shop	53	Whole numbers, decimals, units of measure, accuracy, area, volume, ratio, scales, diagrams.	6, 14, 17, 9, 21, 23, 10, 11, 27
12 Cut2Perfection	55	Whole numbers, accuracy, ratio, fractions, percentages, decimals, tables, pie charts.	6, 9, 10, 12, 15, 14, 24, 25
13 The Pet Shop	57	Whole numbers, accuracy, decimals, percentages, fractions, scales, area, volume, statistics (mean and range), units of measure, diagrams.	6, 9, 14, 15, 12, 16, 11, 21, 23, 32, 17, 27
14 Crafty Wood	59	Whole numbers, units of measure, accuracy, decimals, statistics, percentages, area, ratio, scales, diagrams.	2, 17, 9, 14, 32, 15, 21, 10, 11, 27

TASK 1: SWIFT FEET

Student Information

In this task, you will be converting race time numbers from figures into words.

You will also be calculating a cyclist's and a swimmer's performance by working out mean figures, then creating summary tables.

REMEMBER:

Break the task down into small manageable parts.

Check your calculations at various points.

Make sure your calculations make sense.

Put into your own words what you found.

Writing figures in words

Scenario

Timing is crucial in most sporting events. Think of the men's 100m sprint as an example, where the current world record is 9.78 seconds. When these records are broken it is usually in one hundredth of a second, in other words as little as 0.01 of a second. The difference in time between a world record holder and other athletes can be very small.

Activities

1. A female athlete has just completed a time of 12.96 seconds in the hundred metres hurdles. Write this time in words.

2. The previous record was 12.98 seconds. Work out the difference in time and then write the time difference in words.

3. A top cyclist has just won a three-day event. In the race there was a prize of £65,755 to the winner. Write this figure in words.

4. Over the year the cyclist won a number of European events. The money won came to a total of £482,964. Write this figure in words.

5. The same cyclist has been practising for a 50 kilometre race and recorded his times. Calculate the mean (average) time from the table.

Time in minutes									
74.3	74.1	74.3	73.2	74.1	72.5	73.5	72.1	74.2	72.5

6. Work out the cyclist's range of time from the table above.

7. Place the results of the mean time and range figures into a summary table. Describe your findings.

9 Over a ten-day training period a swimmer has recorded times for a 50m swimming event. The times are placed into the following table:

26.45	26.55	25.69	26.42	25.98
26.74	25.94	26.25	25.85	26.13

Calculate the mean time from the table. Comment on the results.

10 Work out the range of swimming times.

11 Place the results of the mean time and range figures into a summary table. Comment on the results.

TASK 2: STEPPING STONES

Student Information

In this task you will be involved with reading and recording the weight and temperature of some children.

You are also asked to compare the range of weight, height and age of two groups of children registered with the nursery, then choose a suitable way to display your findings.

REMEMBER:

Break the task down into small manageable parts.
Carefully read scales and diagrams.
Convert between units.

Comparing and displaying numbers

Scenario

You are employed as a member of staff at **Stepping Stones** looking after young children.

Childcare and nursery establishments are becoming increasingly popular as mothers tend to return to work soon after having a baby. Today your tasks involve duties you would be expected to complete whilst working in a nursery.

Activities

Complete the following activities that you may have to do whilst caring for young children.

You are required to record information on a child when requested.

1   Look at the scale and find the weight of a child in pounds. Record this weight.

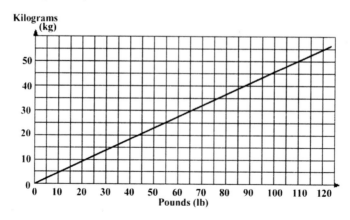

2   There are 14 pounds in one stone (st), change the weight to record in stones and pounds. Describe and clearly show your method of calculation.

3   Another child weighs 55lbs. Using the line graph on *page 38*, plot the weight in pounds and then find the weight in kilograms. Record this information.

4   Sometimes you have to take the temperature of a child for monitoring purposes. Read the scale on the thermometer and record the information.

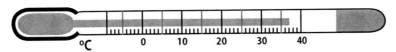

°C   0   10   20   30   40

5   You have to produce a report now. Place the results from the previous recordings in **Activities 1 – 4** into a suitable table. Describe and label your results correctly.

6   Look at the following records of information recorded over one week:

Group A					Group B				
I. D.	Age	Weight kg	Height cm	Sex	I. D.	Age	Weight kg	Height cm	Sex
1	3yrs 6mths	28	91	M	1	2yrs 4mths	16	87	F
2	4yrs 2mths	32	105	M	2	4yrs 3mths	33	103	M
3	3yrs 2mths	29	93	F	3	2yrs 9mths	20	75	M
4	2yrs 9mths	24	74	M	4	2yrs 11mths	24	78	M
5	4yrs 8mths	40	108	F	5	3yrs 1mths	21	81	F
6	3yrs 6mths	31	87	F	6	4yrs 9mths	39	106	M
7	4yrs 7mths	34	102	F	7	4yrs 1mths	38	102	F
8	3yrs 11mths	27	101	M	8	3yrs 3mths	27	88	M
9	3yrs 3mths	22	96	M	9	4yrs 1mths	40	105	M
10	2yrs 11mths	18	78	F	10	2yrs 8mths	18	77	F

Calculate the mean and range of weight in kilograms for Group A and also for Group B. Use an accuracy of one decimal place.

7   Using the previous results, calculate the difference between the mean and range of Group A and Group B.

8   Work out the mean (average) age in months for Group A and the same for Group B.

9   Display all the results from the two groups in a suitable way and highlight the main points of your findings.

TASK 3: DISCOUNTED

Student Information

In this task you will be working out discounts by using percentage calculations in a variety of tasks.

Percentage values are used in everyday items, e.g. discounted clothes. People's wages also include percentages where a tax deduction of a certain value is used or even a percentage bonus is added.

REMEMBER:

Break the task down into small manageable parts.
Use a numberline to see the relationship of percentages.
Show your methods of working out.

Percentages

Scenario

Today you are working in a music shop called **Discounted**. As its name suggests, it specialises in discount prices.

Activities

The retail industry uses percentage values when advertising sale items. These are generally designed to catch a customer's attention and to get them to purchase an item.

The job you are doing now is working out percentage values of sale items.

1   Discounted is having a fantastic sale on music CDs. They are clearing old stock.

   Dark Side of the Sun has got 50% off the regular price of £9.72.

   Work out the discounted price the customers will have to pay.

   Provide an alternative method to find the same answer.

2   The owner still has some old vinyl LPs that are now becoming collectors' items.

   She has placed them on sale at £25.60 each. The sale sign is offering 10% off this price. Calculate the new value of the LPs.

3   The owner has also come across some old cassettes and placed these on sale at £18 each with a discount of 15% of the selling price. How much will she charge for a cassette?

   How much will a customer save if purchasing two cassettes?

4   The owner has a weekend shop assistant who is paid £8 per hour and he works a total of ten hours.

   Calculate his pay and deduct 25% as tax. Show your working out.

5   Another full-time employee has a different tax code and pays less tax on his wages.

If he earned £250 one week and had to have 23% of this deducted as tax, how much will he have to pay in tax that week? Clearly describe and show your method of calculation.

6   The music industry receives royalties on all music sold, which is then given to the musicians. This is calculated as 2% on every item of music sold.

If a new music CD is sold at £12.50, how much will the musician receive as a royalty? Show your method of working out, then show an alternative method of calculation that could give the same answer.

7   VAT is to be added on all items sold and is generally a value of 17.5%.

If a music tee shirt is sold to a customer for £10, how much extra is added as VAT?

Show a method of working this out.

8   There is a sale one weekend for 90s pop music on DVD.

£9.75	£10.25	£13.50	£7.99	£10.85
£12.50	£15.60	£8.99	£12.49	£11.99

Calculate the mean price of the DVDs and also calculate the range. Clearly show your method.

TASK 4: HALF BAKED

Student Information	REMEMBER:
In this task you will be calculating volume, area, the profit ratio of customers, and reading scales.	Break the task down into small manageable parts. Clearly show all methods of working out. Show you are checking your calculations for accuracy.

Volume, weight, check calculations and profit

Scenario

Today you are working in a bistro named **Half Baked**. It is situated in the centre of town and is very popular. You are employed as a chef and occasionally help serve customers.

Activities

1   The ingredients for baking scones are shown here:

Ingredients for 8 scones	
200ml full fat milk	150g sultanas
250g self raising flour	40g butter

Show a method that will calculate how much butter is required to make 64 scones.

2   Cream teas are served at a cost of £2.50 in the bistro and during the afternoon sixty are sold making a profit of £150.

Show a check calculation that would make sure the profit is correct.

3   During one busy weekend the customers using the bistro were in the ratio of 3 : 7 males to females. If a total of 200 people came in, calculate how many males to females there would be.

Clearly show your method of working out.

4   The bistro makes a turnover of £12,672 one week. Write this figure in words.

5   Bread is baked in cooking tins. Calculate the volume of the following bread baking tin in centimetres cubed.

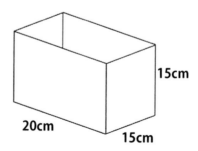

15cm

20cm

15cm

6   Flour is weighed to make bread. Look at the following set of kitchen scales and accurately read the scale to the nearest 1/4 pound.

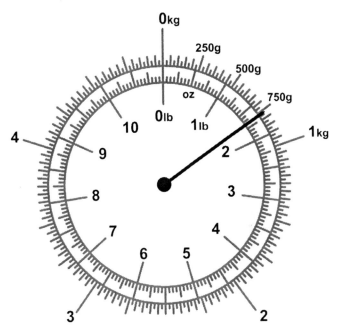

7   New carpet is required for the bistro. Look at the following diagram of the bistro.

Calculate the total area in metres².

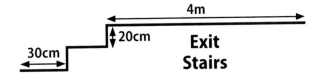

8   If new carpet costs £22.50 per m²; work out the cost of the new carpet for the bistro.

9   A new piece of carpet is also required at the exit stairs of the bistro.

Look at the diagram of a side view of the stairs.

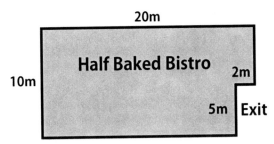

The exit stairs are 2 metres wide. Calculate the area of the stairs.

10   Calculate the additional cost of carpet for the exit stairs.

11   Form a total cost of carpet for the bistro and the area in metres squared covered.

TASK 5: AT A FRACTION

Student Information
In this task you will be becoming familiar with fractions and fraction values. Fractions are used widely, for example, "**Amazing 1/2 Price Sale**" on a poster in a shop window.

REMEMBER:
Break the task down into small manageable parts.
Use a numberline to see the sizes of fractions.
Show your methods of working out.

Using fractions

Scenario
You are working for a firm called **Oxford Flooring** that sells floor coverings ranging from carpets to vinyl flooring. Today your work involves using fractions to work out discounted prices, and using ratio calculations to learn about customers visiting the shop and people attending for interviews.

Activities
The media industry is very competitive and advertisements are carefully worded to attract people into doing something or to purchase something. Your task now is to use fractions and calculate their values.

1   Oxford Flooring has an advertisement on local radio discounting bathroom carpet. The normal price is £9.50 a square metre. It is on sale at 1/2 price. How much is the bathroom carpet now for a square metre?

2   If a customer has bought seven square metres, how much will this cost?

3   The same bathroom carpet is purchased by another customer. She has noticed that there is also a special offer of 1/4 off the normal price of carpet underlay. This normally costs £2 per square metre.

    How much will she pay for five square metres of bathroom carpet and underlay?
    Clearly show your method of working out and show how much she will save in total.

4   Kitchen vinyl normally sells at £9.99, but is now offered at a mega 1/3 off at Oxford flooring.

    A customer has bought seven square metres. What is the new discounted price?

5   An installer charges £1.50 per square metre. What will the customer pay in total?

6   The manager of Oxford Flooring kept a record of sales one Saturday. There were 94 customers who were old age pensioners and this was a 1/5 of the total. How many customers came into the shop?

7   The manager has advertised for a number of jobs at a brand new carpet superstore and is interviewing 15 people.

    A third failed the first interview.

    At the second interview 2/5 failed the practical.

    Calculate how many people got the new jobs. Clearly describe and show your chosen methods of working out.

TASK 6: WHAT'S THE POINT

Student Information

People use decimals daily in items such as currency. In 1971, when the UK changed its currency from the imperial system to the new one it was known as **decimalisation**.

REMEMBER:

Break the task down into small manageable parts.
Use a numberline to see the relationship of decimals.

Working with decimals

Scenario

Engineering is still one of the biggest industries in this country and the use of decimals is extremely important as it allows for precision and tolerances of manufactured components. Today you work for an engineering company manufacturing parts.

Activities

1   A new building frame is being fabricated from steel and the engineer is using a tape to measure a steel girder prior to cutting.

    The engineer measures the length of the steel girder as 8.5 metres. He requires 8.12 metres for manufacturing purposes. How much needs to be cut off in metres?

2   What is this figure to be cut off the girder in centimetres, and in millimetres?

3   There is a plate to be welded at one end of the steel girder. The plate is 25mm thick. What is the final length of the steel girder in centimetres?

4   The new steel girder is to be welded to a new section of structural steel. This piece is 0.355 metres long. Calculate the finished length of the steel girder in metres.

5   What is this new finished length in centimetres and also millimetres?

6   The finished steel girder is weighed prior to transport. Look at the following read out and round the figure off to the nearest whole kilogram:

174.963 Kg

7   A lorry has to transport the finished steel girder to a construction site. The fuel tank on the lorry holds 25 gallons when full. If there are 4.55 litres in one gallon, how many litres of fuel will the tank hold?

8   If fuel costs £0.96 per litre, how much money it will take to fill the fuel tank on the lorry?

9   The manufacturer of the lorry states that the fuel economy is 3 kilometres per 1 litre of fuel. Using the tank capacity in litres, calculate how far in kilometres the lorry will travel. Clearly describe and illustrate your methods of calculation.

TASK 7: DECORATED PAINT

Student Information	**REMEMBER:**
In this task you will be working out the area of a number of rooms and calculating paint ratios.	Break the task down into small manageable parts. Follow instructions carefully. Show all your working out.

Areas

Scenario

You are currently employed by a painting and decorating firm **Decorated Paint**, working as a painter and decorator working in people's homes.

You will be dealing with problems that you may come across as a painter and decorator. The ability to work out area and perimeter is important. Imagine not having enough paint to complete a job.

Usually you have to work from a diagram of a particular job before undertaking it. These are some of the activities that you are to do today.

Activities

One of the first things you may have to do is work out area.

Working out areas is extremely important to a painter and decorator.

1   Look at the following diagram of a room's wall and calculate the area in metres squared using length x width.

      This will aid pricing and the time required for a job.

      Use an accuracy of one decimal place and show your calculation.

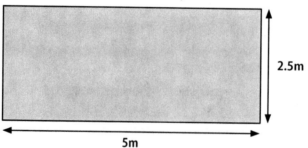

      2.5m

      5m

2   What would the area be if the previous answer was rounded off to the nearest one metre?

3   You are asked to work out the area of another room's wall. This wall has a window in it.

Find the area of the wall to an accuracy of 1 decimal place and draw a diagram of this wall using scale of 2cm to 1m. Dimension and label correctly.

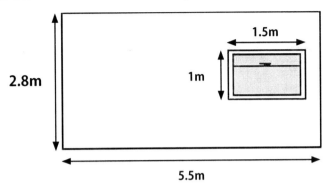

4   Calculate the area of the window in metres squared.

5   Next, take away the area of the window from the area of the room's wall and still use an accuracy of 1 decimal place. Show your working out.

6   Using the previous answer, round off this area now to the nearest whole metre.

7   Occasionally a room may be 'L' shaped and this will require a number of steps to find the area. Find the total area of the following room. Clearly show all of your methods of working out.

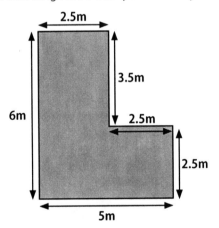

8   Perimeter is another aspect of painting and decorating. For example, you may have to paint a skirting board. Using the diagram of the room above calculate the perimeter of the room. Show all of the working out.

9   Sometimes all the dimensions are not given and this means you will have to work these out.

Using the following diagram of a storeroom, calculate the perimeter of the room. Clearly describe and show your chosen methods.

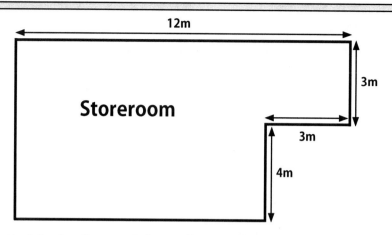

10   Look at the following diagram of a hair and beauty salon.

Calculate the perimeter of the beauty area, manicure/pedicure and hair salon area. Clearly show all your chosen methods of calculation.

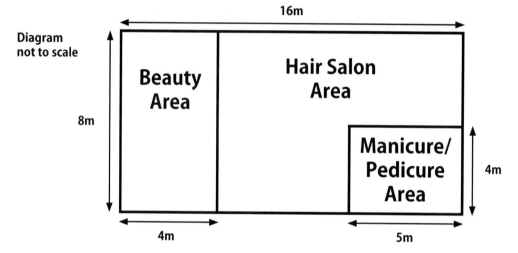

11   Paint is generally applied by using a paint roller and a paint tray. Calculate the volume in centimetres cubed of the following paint tray.

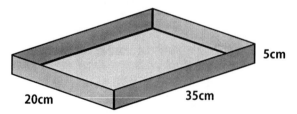

12   A particular shade of green paint is mixed using 2 parts cobalt blue and 1 part signal yellow.

How much paint of each colour is required to make 9 litres of green paint? Clearly describe and show your chosen method.

TASK 8: WELL DISPLAYED

Student Information

In this task you will read information related to activities in a sports club.

There will be information in tables and charts, and you will create tables and charts.

REMEMBER:

Break the task down into small manageable parts.

Carefully follow instructions.

Show you are checking your answers for accuracy.

Describe your working methods.

Presenting information in tables and charts

Scenario

Displaying information can be difficult at the best of times. Nobody really likes looking at just numbers, and displaying a lot of numbers can make reading difficult and even boring.

Displaying data and handling data can be made easier when using the correct method of display. Tables, charts and graphs display data in a way that is easier to understand and it can be possible to compare sets of data in this way.

Your task today is to present numerical data from a sports club.

Activities

Complete the following activities about handling and displaying data:

1   During the month of March you have recorded the number of events that were played by the under twenty-one-year-old teams. The information is displayed in this table:

| Under 21s Sports Events in March ||
Sport	Number
Rugby	24
Cricket	20
Football	12
Basketball	16

You are now going to use this information on sport and display it as a pie chart. Include all angles and correct labels for each sport.

Clearly describe and show your chosen method of working out. Provide a check calculation.

(Continued over)

2   Over the course of one week the number of people who attended the sports club has been recorded and displayed as a horizontal bar chart.

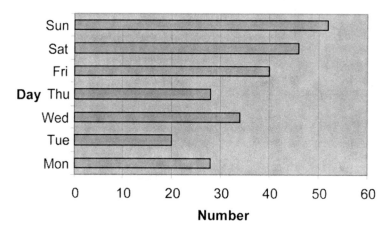

Sports Club Attendance

Using the bar chart, find the most popular day and the number of people who attended on that day.

3   Which is the least popular day and by how many attended on that day?

4   Explain why you think there is a difference between these two days. Evidence this by displaying the difference as a calculation.

5   There is a satellite TV display for club members on which to enjoy major sporting programmes.

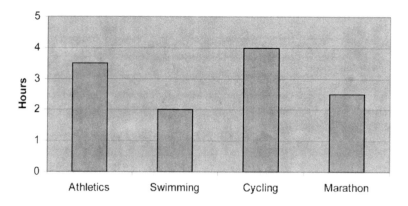

TV Sport

Calculate how many hours in total the sport was on TV?

6   Using the information shown in the bar chart on TV sport, prepare a suitable frequency table.

7   The football club committee has decided to purchase some new football equipment for the Sunday league. Look at the following bank repayment table:

Bank Repayment Period					
Amount of Loan £s	12 Months	24 Months	36 Months	48 Months	60 Months
£10,000	£916.67	£504.17	£369.72	£305.02	£268.42
£7,000	£641.67	£352.92	£258.81	£213.51	£187.89
£6,500	£595.83	£327.71	£240.32	£198.26	£174.47
£6,000	£550.00	£302.50	£221.83	£183.01	£161.05
£5,500	£504.17	£277.29	£203.35	£167.76	£147.63
£5,000	£458.33	£252.08	£184.86	£152.51	£134.21
£4,500	£412.50	£226.88	£166.38	£137.26	£120.79
£4,000	£366.67	£201.67	£147.89	£122.01	£107.37
£3,500	£320.83	£176.46	£129.40	£106.76	£93.95
£3,000	£275.00	£151.25	£110.92	£91.51	£80.53
£2,500	£229.17	£126.04	£92.43	£76.26	£67.10
£2,000	£183.33	£100.83	£73.94	£61.00	£53.68
£1,500	£137.50	£75.63	£55.46	£45.75	£40.26
£1,000	£91.67	£50.42	£36.97	£30.50	£26.84

The football committee requires £4,500 for the new equipment. Look at the repayment table and identify the monthly repayments offered for each repayment period.

8   Place the information into a table that shows five sets of information that includes: the repayment period, monthly repayment and how much will be paid back in total for that period. Clearly describe your findings.

TASK 9: TO SCALE

Student Information

In this task you will become familiar with using scales and proportion. Scales are used in such things as photocopying, road maps and diagrams for manufacturing components used in industry.

REMEMBER:

Break the task down into small manageable parts.

Carefully follow instructions.

Check that you are using the correct units for scaling up or down.

Presenting information to scale

Scenario

Today you are employed in a design office called **To Scale**. Your job will be to read scale diagrams and make calculations related to an engineering part; a house plan; a restaurant extension; and fitting a kitchen.

Activities

1   The military has specified a component for a vehicle. A colleague has produced a drawing of the component. Look at the following scale diagram.

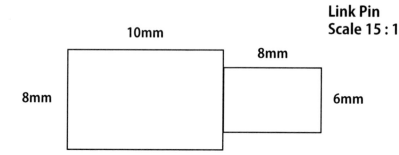

Link Pin
Scale 15 : 1

10mm

8mm

8mm

6mm

Diagram not to scale

Using the scale provided, work out the size of the Link Pin drawn on the diagram. Clearly show your chosen method.

2   A building company has asked you to produce a scale diagram of a house that they build. They have provided you with the plan view of a house (*shown on page 53*).

Draw this house to scale. Use a scale of 2cm to 1m.

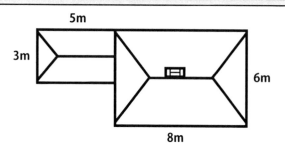

Include the full size dimensions on the diagram.

3 The same building company has recently built a new restaurant extension for a hotel.

Produce a scale diagram of this restaurant using a scale of 1cm : 1m. Include all full size dimensions. Also calculate the restaurant area in metres squared.

4 There are some tables to go inside the restaurant. The dimensions of the tables are 1.5m long by 1m wide. Using the diagram of the restaurant, draw six tables to scale. Placing them in suitable positions.

5 There is also a new kitchen attached to the side of the restaurant. The kitchen fitters are using the following diagram:

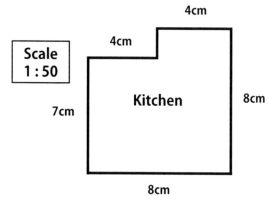

Show a method that will work out the life-size dimension of 8cm.

6 There is also a food preparation table to go into the kitchen. The dimensions of the table are 1.5m by 1m. What size would the table be on the scale diagram?
Clearly show, and describe, your chosen method.

TASK 10: MIXING IT UP

Student Information	REMEMBER:
In this task you will become familiar with the relationship between percentages, fractions and decimals.	Break the task down into small manageable parts. Use a numberline to see the relationship of percentage, fraction and decimal numbers.

Relationships between percentages, fractions and decimals

Scenario

To be effective in employment it is necessary to see the relationship of certain numeracy functions. Percentages, fractions and decimals have a close relationship with each other.

Aerospace Technology is offering a well paid job for which you are applying. Today you are going through the interview process and finally getting the job.

Activities

1. You are one of 30 people who have applied for some new jobs at an Aerospace Technology company.

   50% are rejected initially because of poor qualifications. The remainder make it to the first interview. You are one of the successful ones. How many are there in total?

2. At the next stage 1/3 of these fail the first interview, but again you are successful. How many people fail this interview?

3. At the next and final interview stage 0.8 are unsuccessful and the remainder are appointed in to the new posts. How many people are unsuccessful at the final interview?

You are one of Aerospace Technology's new employees and there is a salary of £25,000 per year for this post. However, this is made up of various bonuses and deductions. For **Activities 4 – 6** clearly show all your chosen methods for each activity. Explain clearly why you chose these methods.

4. Your salary includes a 1/5 bonus scheme for meeting deadline targets. Calculate the value of the bonus.

5. The new job also has a pension scheme. Aerospace Technology contributes the value of 5% on top of the salary. Workout how much aerospace pays in addition to your salary.

6. You have also opted for the health plan. This has a deduction of 0.1 of the salary. Calculate the cost of this health plan.

7. Finally, form a table that shows your salary, how much is made up by a bonus, how much Aerospace Technology contributes as a pension and how much money is deducted for a health plan.

TASK 11: ONE STOP SHOP

Student Information	REMEMBER:
In this task you will calculate volume, dimensions and weight using a variety of measurements.	Break the task down into small manageable parts. Check that you have used the correct units. Show your calculations. Say how you know your results are accurate.

Volumes

Scenario

To be effective in employment it is becoming necessary to become multi-disciplined, in other words capable of doing a number of different jobs.

You are an employee of a construction company called **One Stop Shop** that undertakes various projects for clients. The company employs people who are capable of doing various tasks as and when required.

Activities

1   The first task for today is to do some painting. Look at the following diagram of a can of paint and calculate its volume in cms^3. Clearly show the full calculation.

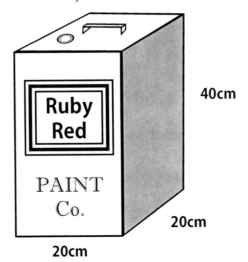

Ruby Red

PAINT Co.

40cm

20cm

20cm

2   Using the previous answer, change the dimensions to metres and give your answer in m^3 for the can of paint. Describe your results.

3   Looking at the instructions on the can of paint, it states that 10,000cm^3 will cover an area 8 metre squared.

How many metres squared will this can of paint cover approximately? Clearly describe and show your chosen method.

4   The next day there is some brickwork to complete. Look at the following dimensions of a brick:

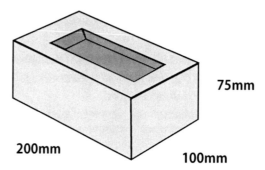

75mm

200mm   100mm

Calculate the volume of the brick in millimetres squared.

5   Using the previous diagram of the brick, convert the dimensions into centimetres and using these calculate the volume of the brick in cm^3.

6   Using the same brick you have been tasked to build a wall for a customer. The dimensions are provided in the diagram. The wall is one brick wide.

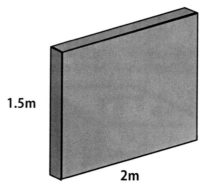

1.5m

2m

Keeping the dimensions of the brick in centimetres, calculate how many bricks will be required to build this wall.

Clearly describe and show your method of calculation.

7   Calculate the volume of the brick wall in centimetres^3. Also calculate the volume of the wall in metres^3.

8   A mortar mix of 3 parts sand to 2 parts cement is required for another job.

What weight of sand and cement is required to make 3.5kg of mortar? Include a check calculation to prove your answer is correct.

9   A work colleague has weighed out some sand prior to mixing with cement. Look at the set of scales opposite and accurately read and record the weight to one decimal place.

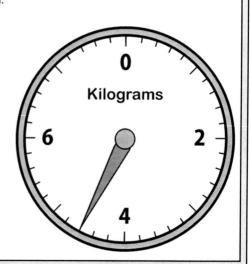

TASK 12: CUT2PERFECTION

Student Information

In business you need to be able to read, analyse and present information in different ways. Here you will analyse tables and graphs, and create a pie chart.

REMEMBER:

Break the task down into small manageable parts.

Clearly show your chosen methods.

Label your work correctly.

Tables and charts

Scenario

The hair industry is becoming more and more popular for both female and male employees. There is also a growing need for people to travel out to customers' homes to carry out treatments.

You are employed at a **Cut2Perfection** salon and have to undertake a number of tasks including monitoring customers' details.

Activities

1   The receptionist has recorded the number of appointments over a week.

Female	Monday	Tuesday	Wednesday	Thursday	Friday	Saturday
Morning	12	14	11	18	22	36
Afternoon	16	21	10	26	26	27
Evening	14	9	19	15	24	18
Home visit	3	4	2	5	8	

Male	Monday	Tuesday	Wednesday	Thursday	Friday	Saturday
Morning	6	4	9	8	10	18
Afternoon	13	11	12	15	18	12
Evening	17	16	15	19	26	6
Home visit		2	1		2	

Form a total for both male and females and then calculate the ratio of females to males.

Clearly show and describe your methods.

2   During a week in May there were 212 female customers having hair treatments. 1/4 of these also had another service. How many were having more than one service?

3   Using the previous answer, express this as a percentage.

4   One busy Saturday there were 72 customers in total and the information was placed into a pie chart for easy comparison.

Hair Colour

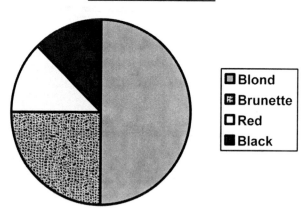

☐	Blond
▒	Brunette
☐	Red
■	Black

Calculate and show how many customers were brunette.

5   From the pie chart, what would be the angle in degrees for the red hair colour?

6   What fraction would be represented by the hair colour black?

7   Hair treatments have been recorded over a week. You are to draw a pie chart to show this information of customers to treatment type.

Cost	Treatment Type	Customer Numbers
£18	Cut 'n' blow	24
£25	Dye	18
£35	Extensions	12
£30	Highlights	20

Clearly describe your chosen method and correct any mistakes. Display the segments in degrees and label correctly.

8   The salon is replacing its old stock with brand new equipment. You are to borrow £7,000.

Study the bank repayment table:

Bank Repayment Period					
60 Months	£268.42	£187.89	£174.47	£161.05	£147.63
48 Months	£305.02	£213.51	£198.26	£183.01	£167.76
36 Months	£369.72	£258.81	£240.32	£221.83	£203.35
24 Months	£504.17	£352.92	£327.71	£302.50	£277.29
12 Months	£916.67	£641.67	£595.83	£550.00	£504.17
Loan £s	**£10,000**	**£7,000**	**£6,500**	**£6,000**	**£5,500**

How much money will the salon pay back on a 36 month repayment plan?

9   Form a total amount that the salon will pay back. Clearly show your chosen methods.

10   Calculate how much profit the bank will have made, after deducting the original loan.

TASK 13: THE PET SHOP

Student Information

In this task you will work out the hours employees have worked and calculate wages.

You will perform volume and area calculations.

You will interpret information contained in a diagram and calculate travelling distances.

REMEMBER:

Break the task down into small manageable parts.

Check your calculations at various points.

Describe your methods of calculation.

Decimals, percentages, volume, area and distance

Scenario

Today you are the owner of a pet shop. It requires working long hours. You also have staff working with you in the shop. You will look at wages and a delivery of goods.

Activities

1   Over one week you have to record hours worked by members of staff prior to working out their wage. You have placed the hours of one salesperson into the following table:

Mon	Tue	Wed	Thurs	Fri	Sat	Sun
8hrs	6hrs	5hrs	10hrs	8hrs	5hrs	7hrs

Form a total for the hours worked. Calculate the mean (average) hours worked over the seven days and also the range of hours worked.

2   The salesperson is paid £5.80 per hour Monday to Friday. Saturday is paid at £5.80 x 1.25 per hour worked, and Sunday is paid at £5.80 x 1 1/2 per hour worked. Calculate the final wage for the week. Show your methods of working out.

3   Tax is deducted from the final wage and is based on 25%. Deduct 25% as tax from the final wage and give an answer. Round off the wage to two decimal places to allow for currency.

A customer who owns a kennel has requested a large delivery of cans of dog food.

The dimensions of a can are shown below:

10cm

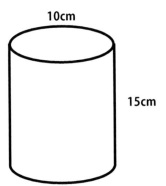

15cm

The cans are to be placed into this large packing box for delivery.

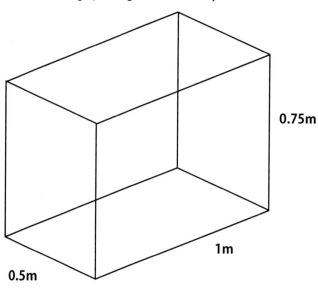

0.75m

1m

0.5m

4   Work out the volume of the packing box in cubic metres.

5   Calculate how many cans of dog food can be placed inside the packing box.

Clearly describe and show your chosen method.

6   The customer has asked for the dog food to be delivered to her kennel business.

The shop has a delivery van. However, the van driver needs to put some petrol in the van from the local petrol station on the way.

Look at the following map and calculate how far the van driver has travelled.

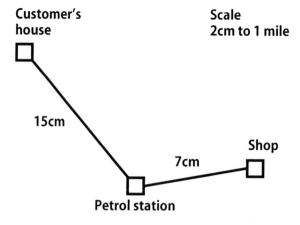

Customer's house

Scale 2cm to 1 mile

15cm

Shop

7cm

Petrol station

TASK 14: CRAFTY WOOD

Student Information	**REMEMBER:**
In this task, you will be interpreting a range of plans, diagrams and tables and calculating dimensions, using a variety of measurements.	Break the task down into small manageable parts. Find missing dimensions. Use correct units of measure. Select suitable ways of presenting your findings. Describe your results.

Dimensions

Scenario

Traditional construction trades such as joinery are becoming increasingly popular areas of employment. Today you are employed in a construction business called **Crafty Wood** and working for a customer who has requested some carpentry work to be done in their house and garden.

Activities

You are to undertake some jobs at a customer's house. Look at the following plan view (overhead) of the house and garden.

A new wooden fence is required around the property.

1   From the dimensions given, calculate the total perimeter of the garden to find the length of the fencing material required in metres.

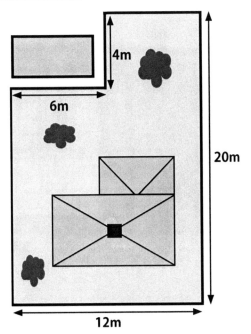

2   The customer has asked for a number of price quotations for different types of fencing.

Look at the following table on fence panels and calculate the price for each fence type and select an effective way to show these calculations:

Crafty Wood		
Fence type	Height	Cost per metre
Highland Lattice Panel	1m	£15.45
Dome Board Panel	1.5m	£21.99
Longwell Lap Panel	2m	£25.50
Vertical Spruce Panel	1m	£18.25
Crosshatch Pine Panel	1.5m	£20.99

Show your calculations and check the answers with a calculator.

3   The customer has decided to purchase the Crosshatch Pine Panel fence. Calculate the difference in cost of the Crosshatch Pine fence compared against the cheapest fence. Clearly show the difference, and show the price range of available fence types.

4   A neighbour of the customer has asked for a quotation. However, there has been an increase of 10% because of a rise in material and labour costs. How much does the neighbours' quotation amount to?

5   The original customer's entrance to the house needs a new replacement door and door frame. Look at the following diagram:

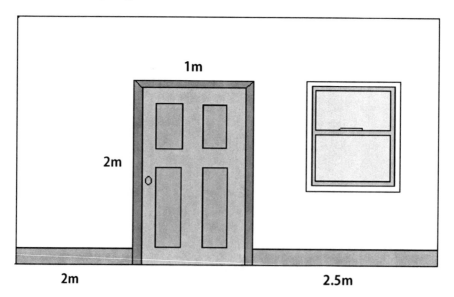

1m

2m

2m

2.5m

Diagram not to scale

If the door frame is 50mm wide, what is the size of the door required? Give your answer in metres squared. Clearly describe and show your chosen method.

6   Convert the previous answer and show the dimensions in centimetres squared.

7   The garage roof has dry rot and requires replacement joists.

Using the following scale drawing, calculate the full length of the horizontal and angled joists in centimetres. Clearly show and describe your chosen method.

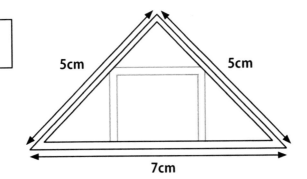

8   Convert and show the answer of the horizontal and angled joists in metres.

SAMPLE END ASSESSMENT

20 Multiple-choice questions

The following questions are multiple-choice. There is only one correct answer to each question.

Instructions

1   Choose whether you think the answer is A, B, C or D.

2   Ask your tutor for a copy of the answer grid (or download a copy from **www.lexden-publishing.co.uk/keyskills**).

3   Enter your answer on the marking grid at the end of the test.

4   Hand it to your tutor for marking.

An Application of Number Key Skills Level 1 External Assessment will consist of 40 questions and you will have **1 hour and 15 minutes** to complete them.

How will you select your answers?

If you are sitting your End Assessment in paper format – not doing an online test – you will have to select one lettered answer for each numbered question. The answer sheet will be set in a similar way to the example below:

1   [a] [b] [c] [d]

2   [a] [b] [c] [d]

Make your choice by putting a **horizontal line** through the letter you think corresponds with the correct answer.

Use a pencil so you can alter your answer if you wish and take an eraser to allow you to change your mind about a response. Use an **HB pencil**, which is easier to erase. (If you make two responses for any one question, the question will be electronically marked as **incorrect**.)

Take a **black pen** into the exam room because you will have to sign the answer sheet.

Your tutor has 100 sample End Assessment questions and you will be given these when your tutor considers you are ready to practise the questions.

QUESTIONS

1   There are sixty-seven thousand one hundred and forty-nine people watching an athletic event. What is this in **written** figures?

A   67,149

B   6,749

C   670,149

D   607,149

2   There are 46,509 people watching a rugby match. What is this to the nearest **thousand**?

A   46,000

B   40,000

C   47,000

D   45,000

3   A builder is making a new patio. He is using flagstones 75cm by 75cm. How **many** flagstones will he require to form a patio 7.5m by 3.75m?

A   15

B   45

C   50

D   75

4 A gardener records the temperature in degrees Celsius in his greenhouse over a 28-day period of time.

Week 1	2	-1	-4	0	2	1	3
Week 2	3	-2	1	-3	0	1	-2
Week 3	5	3	2	7	2	-3	5
Week 4	1	2	-3	2	5	8	6

What is the **range** of temperature?

A 6°

B 8°

C 10°

D 12°

5 If the decimal 0.75 is 3/4 as a fraction, what would it be as a **percentage** value?

A 75%

B 7.5%

C 3.5%

D 3.75%

6 A builder has drawn a diagram of the cavity in part of a loft prior to insulating it:

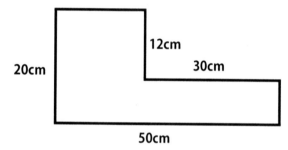

What is the correct **perimeter** of the cavity space prior to insulating?

A 140 cm

B 155 cm

C 160 cm

D 178 cm

7 To hire a car costs £15 per day for the first 5 days, and then £12.50 per day after 5 days. Which of the following shows the correct **calculation** for 10 days' hire?

A (£15 x 5) + (£12.50 x 5)

B (£15 + £12.50) x 5 x 2

C (10 x £15 + £12.50)

D (£12.50 + 5) x (£15 + 5)

8 There were 7,200 people who ran in a London marathon. 1,800 of these were under 21 years old.

What **angle** would the under 21s be on a pie chart?

A 45°

B 30°

C 90°

D 120°

9 Antonio is cooking a batch of pancakes for a restaurant. He uses the following recipe to make 10 pancakes:

Flour Plain 200g

Salt 1 tsp

Sugar 75g

Butter 40g

Milk 300ml

How much butter will he require to make **25** pancakes?

A 85g

B 75g

C 90g

D 100g

10 Markus has brought home €350 from his holiday. If the exchange rate is £1 = €1.40 how **much** will he get if he changes his Euros back in to Pounds?

A £250

B £260.75

C £275

D £280

11 The salesman at a car show room has recorded his sales and placed the information in a chart.

Car registration number	Total cars sold
01	18
51	12
02	20
52	22

If he was to draw a pie chart to show this information, what **angle** would represent the 01 number plate sales?

A   72°

B   90°

C   108°

D   120°

12 A sprinter takes 11.43 seconds to run 100 metres. What is this to the nearest **tenth** of a second?

A   11s

B   11.4s

C   11.41s

D   11.5s

13 An aquarium shop has a sale on fish tanks.

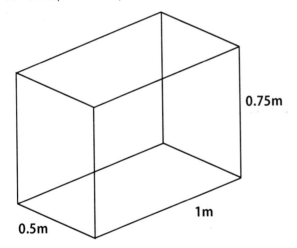

0.75m

1m

0.5m

What is the **volume** of the tank?

A   0.375m³

B   3.75 m³

C   1.375 m³

D   0.75 m³

14 An estate agency charges 2% commission on house sales. How much **commission** will it make on a property sold for £125,000?

A   £250

B   £125

C   £1,250

D   £2,500

15 A machine in an engineering company can produce 240 components in 8 minutes. How many will be **made** in 1 hour?

A   1,800

B   14,400

C   1,920

D   480

16 A swimmer is training for a competition. She swims 1.5km. If the pool is 50m in length how many **lengths** did she swim?

A   300

B   30

C   75

D   175

17 The diagram shows the distance on a digital read out:

2.656 Km

What is the reading to the nearest **10** metres?

A   2,660m

B   2,660m

C   2,600m

D   2,700m

18 A sports club keeps a record of attendances of members:

	Mon	Tue	Wed	Thurs	Fri	Sat	Sun
Adult	12	14	20	22	15	22	35

What is the **mean** attendance figure for adults?

A   26

B   24.5

C   20

D   20.5

19   Look at the following map:

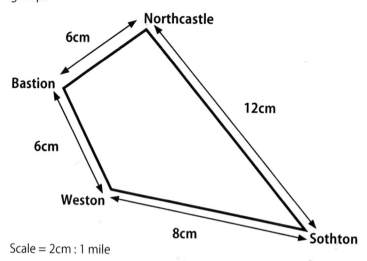

Scale = 2cm : 1 mile

What is the distance in **miles** from Northcastle to Sothton and then to Weston?

A   8.5 miles

B   13 miles

C   10 miles

D   14.5 miles

20   Look at the plan view of a restaurant:

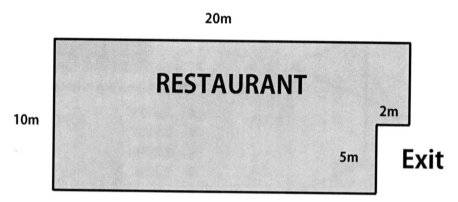

What is the **area** of the restaurant?

A   200m²

B   190m²

C   180m²

D   160m²

66

INDEX

Printed in the United Kingdom by
Lightning Source UK Ltd., Milton Keynes
140559UK00001B/26/P